The History, Progress and Comparative Merits of the Hereford Breed of Cattle

by T. Duckham

with an introduction by Jackson Chambers

This work contains material that was originally published in 1865.

This publication is within the Public Domain.

This edition is reprinted for educational purposes
and in accordance with all applicable Federal Laws.

Introduction Copyright 2018 by Jackson Chambers

Self Reliance Books

Get more historic titles on animal and stock breeding, gardening and old fashioned skills by visiting us at:

http://selfreliancebooks.blogspot.com/

Introduction

I am pleased to present another title in the "Cattle" series.

The work is in the Public Domain and is re-printed here in accordance with Federal Laws.

As with all reprinted books of this age that are intended to perfectly reproduce the original edition, considerable pains and effort had to be undertaken to correct fading and sometimes outright damage to existing proofs of this title. At times, this task is quite monumental, requiring an almost total "rebuilding" of some pages from digital proofs of multiple copies. Despite this, imperfections still sometimes exist in the final proof and may detract from the visual appearance of the text.

I hope you enjoy reading this book as much as I enjoyed making it available to readers again.

Jackson Chambers

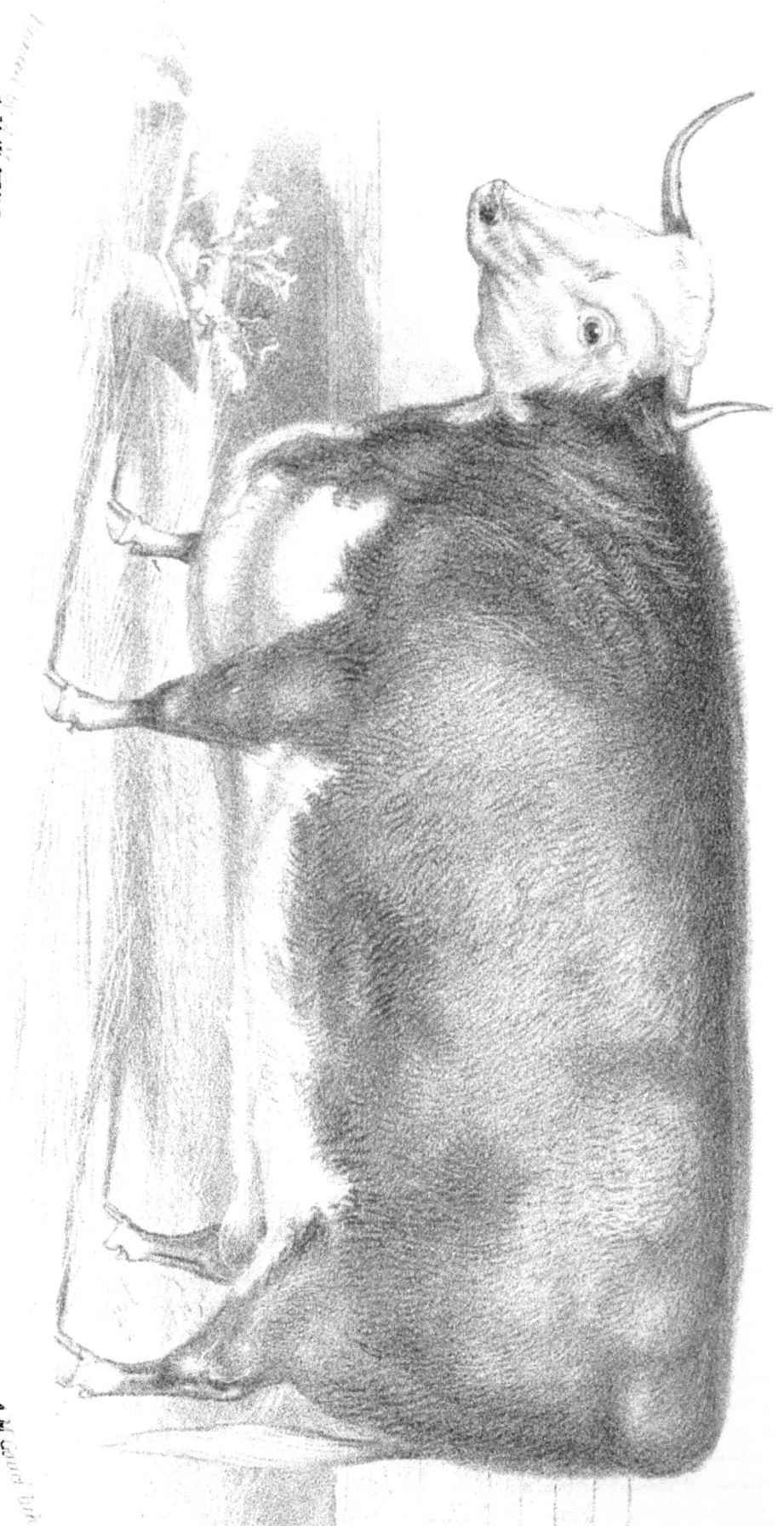

HEREFORD OX at 4 YEARS OLD.

Winner of the Gold Medal as the best Ox or Steer of any breed, at the meeting of Smithfield Club, 1863.
Bred by Mr T. L. Meire, Cound Arbour, Salop, fed by Mr Heath, Ludham Hall, Norfolk.

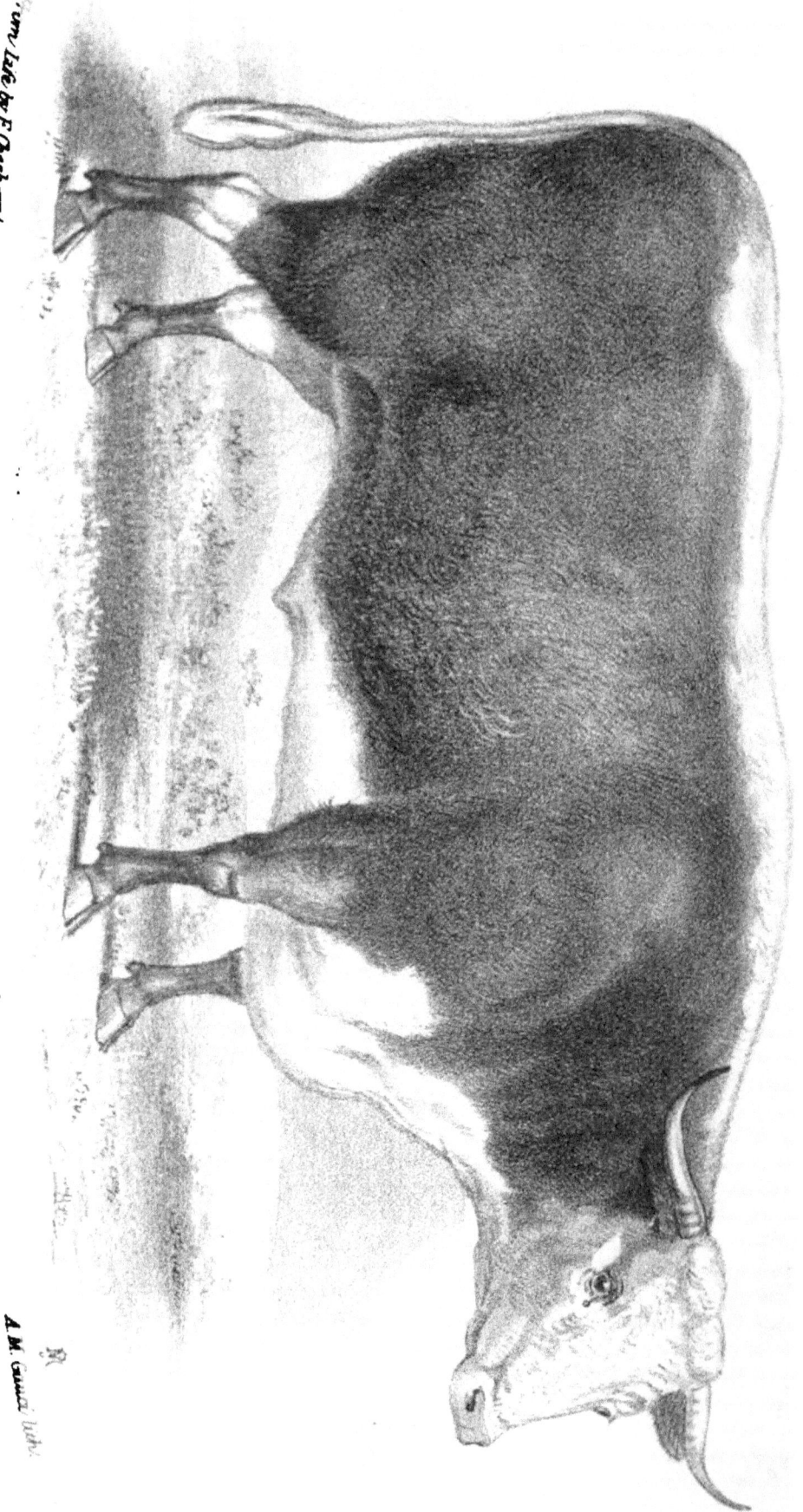

HEREFORD OX at 7 YEARS OLD.
Winner of the first prize at the first meeting of the Smithfield Club, 1799.
Bred by Mr Tully, Huntington Court, Hereford, fed by Mr Westcar, Aylesbury, Bucks.

LECTURE.

Gentlemen,

My appearing before you to day is in a capacity quite new to me. Yet when solicited to do so, I readily complied, from a consideration that the objects of this great national institution should meet with every aid which can possibly be given to it. Actuated by that feeling, although totally unaccustomed to the task, I determined upon adding my mite of information to the great store of valuable scientific and practical knowledge which has long been taught within these walls by gentlemen of high scientific attainments, and through whose unwearied researches many of the previously hidden laws of nature have been brought to light, and who have by their elaborate and lucid explanations of those laws rendered them familiar to you. Still the noble pursuit which you have selected for your after-life is one which knows no bounds to the store of information connected therewith; for as nature herself displays a soil as varied as the climate which regulates the husbandman in its cultivation, so does that soil in its turn become clothed with herbage equally varied: thus as the plants which flourish in the vale are lost upon the hill-top, so will those who hope to profit from her, unless they make her laws their study in the cultivation of the soil and the adaptation of the stock to the position in which it may please the Almighty Ruler of all things to place them.

With this brief introduction I will now endeavour to place before you the result of my researches respecting that beautiful, hardy, and flesh-forming race of cattle known as the Herefords; taking their name from that of the county which constitutes the seat of the breed. That they are an aboriginal race is, I consider, beyond question; and although indigenous to the soil of the county, yet I think before I close my remarks to you this day I shall be enabled to make it manifest that sound judgment is all that is requisite to insure success to those who breed them, not only throughout this kingdom, but also over a large area of the known world.

I am quite aware that upon this point a contrary opinion is entertained by some; but when I read to you the experience of numerous practical farmers whose success or failure in business is dependent upon the soundness of judgment they display in the pursuit of their occupation, I shall unmistakably prove it to be a mere delusion, or, if experienced, to be caused by a lack of judgment on their part rather than from any deficiency in the animals themselves.

As regards their early history little is known, or can be gleaned previous to the establishment of the great national fat-show of the Smithfield Club. But Mr. Rowlandson, in his prize report on the "Farming of Herefordshrie," which appeared in the Journal of the Royal Agricultural Society of England, Vol. xxxii., p. 450, advances an opinion that they were originally brown, or reddish-brown, and insinuates that they originally came from Normandy or from Devon. He also relates the following story which he had heard respecting the appearance of a white-faced bull in the herd of Mr. Tully, Huntington, near Hereford: "About the middle of the last century, the cowman came to the house announcing as a remarkable fact that the favourite cow had produced a white-faced bull-calf. This had never been known to have occurred before; and as a curiosity it was agreed that the animal should be kept and reared as a future sire;" and adds, "that the progeny of this very bull became celebrated for white faces."

The same authority gives an interesting extract from history, showing that in the tenth century a celebrated breed of white cattle with red ears prevailed in Wales, of which that part of the county of Hereford on the north side of the river Wye formed a portion. He tells us that a law of Howell the Good fixed compensation to be paid for injuries done by one of the princes towards another at one hundred white cows with red ears, and a bull of the same colour; and if the cattle were of a dark or black colour, then one hundred and fifty in number instead of a hundred, and adds, "Speed records, that Maud de Brehos, in order to appease King John, who was highly incensed against her husband, made a present to the Queen of four hundred cows and one bull from Brecknockshire, all white with red ears." These facts he says "are suggestive of the mode in which the white-faced cattle have originated."

An old and much-respected friend of mine, the late Mr. Welles, also entertained the idea that they were originally self-coloured like the Sussex or Devon, and "that the breed characterized as the mottle-faced took its origin from a mixture of the old self-coloured with some accidentally possessing white marks;" whilst Mr. T. A. Knight says, "Lord Scudamore, who died in 1671, introduced cows of the red-with-white-face breed from Flanders." Now be all this good for what it may, if the breed of the county were ever self-coloured they have long since passed away. As regards the white cows with red ears I think the light grey or white Hereford, to which I shall presently refer, may fairly be considered to be descended from them; and notwithstanding the opinions I have quoted, there are red-with-white-face breeders who advance that they can trace them as being the breed of their ancestors for the past two hundred years. But as I intend to confine my remarks to you to facts which can be proved, I shall not enter further into any of the surmises which have from time to time been written respecting them, and the probable effects produced by a commingling of blood of the different classes. I shall therefore commence my notice of them with that of the first meeting of the Smithfield Club, which took place in the year 1799. That they were then highly esteemed by the grazier is clear from the fact that the first prize at that meeting was won by a Hereford ox, exhibited by one of the founders of the Club, namely, the late Mr. Westcar, of Creslow, Buckinghamshire, in proof of whose eminence as a feeder I make the following extract from a letter, written by Mr. Arnsby, and which appeared in *Bell's Weekly Messenger*, in May, 1857:—"Mr. Westcar took the first prize with a Hereford ox for twenty years in succession in the London Cattle Show, which was open to all kinds of cattle against Mr. Westcar." (At that time all breeds showed in competition with each other, and not, as now, in distinct classes). Mr. Arnsby also tells us in that letter, that "about the year 1812 or 1813, I saw Mr. Thomas Potter sell for Mr. Westcar fifty Hereford oxen in Christmas cattle market, that averaged 50 guineas each, making 2,500 guineas;" and I have, through the kindness of Mr. Smythies, of Marlow, Salop, obtained the following extract from Mr. Westcar's book of the sale of twenty Hereford oxen at different periods from 1799 to 1811. The gentleman who made it writes thus:—"I have confined myself to such only that sold for £100 and upwards; had I descended to £80 I know not to what extent in number of animals my list would have run.

"1799, Dec. 16,	Two oxen to	Mr. Chapman	£200
1800, Dec. 4,	One do.	do.	147
" " 13,	One do.	Mr. Harrington	100
1801, Nov. 26,	Six do.	Messrs. Giblett & Co.	630
1802, " "	One do.	do.	100
" " 31,	One do.	Mr. Chapman	126
" Dec. 4,	Two do.	Mr. Horwood	200
1803, " "	One do.	Mr. Chapman	100
" " 19,	One do.	Mr. Reynolds	105
" " "	One do.	Messrs. Giblett	105
1804, Dec. 5,	One do.	do.	105
1805, " 4,	One do.	do.	100
1811, Nov. 28,	One do.	Mr. Chandler	105 ;"

thus the twenty oxen realized the sum of £2,123, being an average of £106 6s. each. Mr. Westcar was an ardent admirer of the Herefords; and by his regular attendance at the Hereford October fair, from 1779 to 1819, at which he made not only many purchases, but also induced the Duke of Bedford, and many other noble Lords, to adopt the same plan, he did much to bring this valuable race of animals into public notice at an early period of agricultural progressio

In that extremely interesting and valuable history of the Smithfield Club, compiled with great care, and published by their indefatigable Honorary Secretary, Mr. B. T. Brandreth Gibbs, there is a fund of information given to the world which very probably would otherwise never have been made known. From it we learn that at their first show Mr. Westcar's prize ox measured 8 feet 11 inches long, 6 feet 7 inches high, 10 feet 4 inches girth, and that he was sold for a hundred guineas; and from the dimensions given upon this coloured print, together with the names of the feeder and purchaser, all corresponding with the particulars I have just read to you, I have no doubt it is intended to represent that identical animal; if so, he was bred by Mr. Tully, Huntington, near Hereford: his weight was 247 stones; and although not of the form we should expect to see upon entering Islington Hall next week, yet it is an interesting picture, inasmuch as the distinctive marks of the red with white face of the present day are here set forth, with the exception of the white stripe which now extends along the neck and just over the shoulders, being here shown as far as the hip bones, and also the lower part of the legs being red instead of white. Enormous as the dimensions of that animal were, they were thoroughly eclipsed by another Hereford ox, fed by Mr. Grace, from Bucks; he is said to have measured 7 feet high, 12 feet 4 inches girth, and to have weighed 260 stones. It appears that although we have this interesting information given of the first meeting, yet the records of not only that meeting, but also those of 1801, 1802, 1804, and 1806 were incomplete; and therefore Mr. Gibbs, in a tabular statement which he has furnished of the award of premiums from the formation of the Club until the year 1851 does not give any account for those years. But he tells us on page 20, that "the winners at the first meeting were Mr. Westcar, the Duke of Bedford, Mr. Edmunds, and Mr. John Ellman, the latter for the best ox fattened with grass and hay only in the shortest time from the yoke." Of Mr. Westcar's he had previously told us it was a Hereford; and as the Duke of Bedford and Mr. Ellman were both Hereford breeders, it is only a fair inference to draw that three if not the four of those were winners with Herefords. In the notes of the meeting for 1812 it is remarked that there was "a Hereford with a red ring round his eye," and "a smooth coated Hereford." Now although those remarks appear to be of a trivial nature, and only mentioned by Mr. Gibbs as a curiosity, yet they go far to show that in those days they expected to see a Hereford with a white face and a rough coat at the Smithfield Show. In 1825 he gives a little incident worthy of notice, namely, "a sweepstakes between three Herefords belonging to the Duke of Bedford, and three Durhams belonging to the Right Hon. Charles Arbuthnot, which was won by the Herefords (of course you are aware that the Durham of that day was the shorthorn of the present). From the establishment of the Club to the year 1851 all the different breeds and crossbreeds were shown in competition with each other; since the latter period they have been exhibited in distinct classes. It is to be regretted that the tabular statement to which I have before alluded is incomplete; but as far as we can learn from it during the time they were shown in competition, the Hereford oxen and steers won 185 prizes, the shorthorns 82, the Devons 44, the Scotch 43, the Sussex 9, the longhorns 4, and crossbreeds 8; thus showing that the whole of the prizes won by all the other breeds and crossbreeds in the kingdom were 190, or only 5 in excess of the number registered as won by the Herefords; and as I have before shown, they can fairly lay claim to have been the winners of the majority (if not all) the prizes at the first meeting, whatever may have been their doings at the subsequent years of deficient information. I think it not too much to say, that they did, during the period they showed in competition, equal in number of prizes all the pure breeds and cross breeds in the kingdom; this in itself cannot fail to prove their superiority as a flesh-producing race of cattle. Notwithstanding the success which attended the exhibition of oxen and steers, yet the same tabular statement shows a great falling off with that of the cows and heifers; as only 22 prizes stand to their credit, whilst the shorthorns won 92, the Devons 4, the Sussex 3, the longhorns 6, and crossbreeds 6. This is certainly a great falling off compared with the oxen and steers; and goes far to prove the correctness of my remark respecting the study of nature's laws in the cultivation of the soil, and of the adaptation of stock to it. The soil of the county of Hereford being neither applicable for dairy or feeding purposes, those who have cultivated it have for ages made it their study to breed steers and oxen which should by their superior quality and aptitude to fatten command the attention of the distant grazier. The success with which they have done this for many years past I have shown you; and the demand which continues to exist, proves that there is no falling off in the superiority of their animals for the graziers, and perhaps there is no finer sight for the admirers of cattle than the annual October fair at Hereford, which takes place on the third Wednesday and the preceding day of that month (and not as heretofore on the 20th). On those days several thousands of steers pass from their breeders to the graziers, occupying the fertile pastures of Bucks, Northampton, Kent, Essex, &c.; and whatever may have been their original colour and distinctive marks in days of yore, their present uniform appearance cannot fail to impress those who attend that fair for the first time with a degree of surprise and admiration in their walk through the streets of the city, to see line after line of them all displaying a similarity of character, and at once claiming each other as one family.

The names of Tully, Walker, Tomkins, Jones, Hewer, Yeomans, Yarworth, Weyman, Jeffries, Haywood, Knight, Price, Morris, Skyrme, Williams, Ridgeway, and Griffiths, may be considered amongst the most eminent of their breeders at the commencement of the present century.

There was then no herd-book to guide them in the selection of their breeding animals; there was no great national society which by its annual monster gatherings could concentrate animals of the different breeds in the same show-yard, and whereby comparisons could be made; and thus the breeders frequently resorted to challenges in order to test their superiority. Several of these and their results are recorded in Vols. I. and II. of "Eyton's Herd-Book of Hereford Cattle;" one of them I will here notice, namely, a challenge given in 1810 by Mr. Meek, of Lichfield, to show his bull against any Hereford bull for a hundred guineas. His challenge was accepted by Mr. Walker, Burton Court, who sent his bull, Crickneck (175), to Lichfield; but when he got there it appeared that Mr. Meek had by some means made himself acquainted with the superiority of Mr. Walker's bull, and rather than submit to defeat he allowed judgment to go by default. As I purpose doing myself the pleasure of adding a set of the Herd-Book to your library, I will not here relate any more of those triumphs; but by a perusal of its pages you will see other triumphs mentioned, as also many awards of premiums given to Hereford breeding animals when shown in competition with those of other breeds. The first volume bears date 1845. At that period Mr. T. C. Eyton, of Eyton Hall, Salop, feeling that the want of such a work was a great loss to the breeders, most praiseworthily and assiduously commenced the task of obtaining information, in order that he might be enabled to publish a work he felt to be so much required; but the breeders were quite unacquainted with its value: some looked upon it with a degree of jealousy fearing that if carried out it would show too much of the system they pursued in breeding; others were sceptical of its value; whilst others felt that their particular breed was the best, and should therefore have a herd-book of its own: and so strong was the latter feeling entertained that at one time a decided determination of having one was expressed; and the only way in which it could be appeased was by giving precedence in the first volume to their class of animals, although very greatly to the disarrangement of the work, which decidedly should have been alphabetically arranged from its commencement, whereas now we have "Leopold (1)," "Waxy (3)," "Aaron (82)," and so on; but not only did this precedence of their particular blood (the mottle-faced) disarrange the work, but it also gave dire offence to many of the red-with-white-face breeders. Mr. Eyton failing to obtain sufficient information to satisfy him, he added, as an appendix to the work, several catalogues of sales, giving the names of the purchasers, and prices of the animals where he could obtain the information. That appendix I have found of considerable value in my researches, as it contains the names of many valuable cows; and as that important part of all herd-books, the cows and their produce, was not comprised in Vols. I. and II., I have frequently found it requisite to refer to the appendix for information. The second volume was published in three parts, and completed in January, 1853, when Mr. Eyton announced, "It is not my intention to carry the Work on further unless the breeders generally come forward to assist me more than they have done up to the present time;" adding, "I would willingly give my own time and trouble, if I thought that all would join in working out the truth, and afford the work sufficient patronage to cover its expenses." Strong as this appeal was, it did not arouse them from their apathy, notwithstanding he at the same time told them that several of the volumes had gone abroad. Yet they could not be induced to see that their success was identical with that of the work; and thus, although he made their entries free of charge, and only charged such a sum for each volume as he considered would cover the expenses of a work which necessarily must be one of a limited circulation, but few volumes comparatively were sold; and I have no doubt it would have ceased to exist had not the late Mr. W. Styles Powell, of Hereford, yielded to the solicitations of some few breeders who felt its continuance to be of importance, and therefore urged him to undertake it. In 1856 he purchased the copyright; and in the autumn of that year published the first part of a third volume, containing 236 pedigrees: he then made known his intention of publishing another part the next year; and expressed a wish to add the pedigrees of cows with their produce to that volume if he could induce the breeders to supply him with the requisite information; but although then in the full health and vigour of manhood, a few short months only were allowed to pass before his earthly career terminated, and the existence of the work again appeared in jeopardy. After his decease, his uncle, your respected vicar, and his representative, sold the copyright for a nominal sum to the Committee of the Herefordshire Agricultural Society, who transferred it to me. At that time no attempt had been made to make it a subscription work; and but few of either Mr. Eyton's volumes, or the part by Mr. Powell, were sold: the latter I revised and reprinted, which, with a second part for bulls, and a third part for cows with their produce, formed the third volume, published in the autumn of 1858, under the patronage of that noble patron of all good works, His late Royal Highness the Prince Consort, whose memory will be ever cherished by every true lover of his country's greatness; added to that, however, I was also enabled to publish a List of 187 Subscribers; and knowing that many regretted they had not more promptly responded to my call for information, I resolved to give them another opportunity, and therefore published the fourth volume the following year, with a List of 247 Subscribers; and having previously resolved upon making it a triennial publication, the fifth volume was published last year: the total number of bulls now entered are (2364), and the List of Subscribers 317. At the request of the breeders I have now resolved to publish it every other year, and therefore purpose, should I obtain a sufficient number of pedigrees, to publish the sixth volume next year. The steady increase of public support which I have gained cannot fail to be gratifying to me; and I trust I may ever continue to merit that confidence which it assures me I have obtained. The importance of the work is daily becoming more and more understood by the breeders; and that feeling of secrecy to which I alluded in the commencement of my notice of it, has not only ceased to exist, but every effort is now made to obtain and supply the most complete information possible. Yet that it did exist I think the following little anecdote will go far to show. A farmer having some years ago purchased a young bull with an eminent breeder, sent his son for him, and gave him strict orders to ask for his pedigree. True to his parent's commands he did so; but the only answer he could obtain was, "Tell your father his pedigree is on his back," meaning that the bull had a good broad back, well covered with superior flesh, which should in itself be a sufficient guarantee for his use, at the same time jealously guarding as a profound secret how he was bred. The success which subsequently attended that bull's appearance at the show of the Royal Agricultural Society of England, and many local gatherings, proved his excellence, and supplied the more cause for regret at the narrow-mindedness of his breeder.

When Mr. Eyton commenced his labours with the Herd-Book, he found it requisite to divide the Herefords into four distinct classes, namely, the mottle-face, the dark-grey, the light-grey or white, and the red with white face. Yet, after the lapse of only eighteen years, the all but universal appearance of the red-with-white-face Hereford is such, that when any animals of either of the other classes are exhibited, the purity of their blood is questioned by those who are not cognizant of these facts. Indeed, so far back as 1857, the *Mark Lane Express*, in its report of the Birmingham fat show, asks, "Is there such a thing as a white Hereford? There was one entered and shown as such, though we rather question if he claimed kindred here whether he would have that claim allowed." This query was answered by myself in the affirmative. The present uniformity of colour and marks is due to the influence of the bull; and I think goes far to prove it to be the original breed, let the other classes have sprung from whatsoever accidental or other causes they may.

All the classes are of the middlehorn tribe; and so closely do they assimilate to each other in their principal features, that they are recognized as equally pure-bred when entering the arena of competition. In Vol. I. you will see a coloured engraving of a choice specimen of each class, which will more readily convey to your minds their character than any description in words which I can place before you.

The mottle-face, as the name denotes, had red marks intermixed with the parts usually white, namely, the face, feet, &c.; occasionally those spots were of a very dark colour, the horn was long and wavy, with a slight upward tendency, and tipped with black; their skin was particularly mellow, of a moderate thickness, and well covered with plenty of soft glossy hair; they were not usually good upon the chine; and although not generally so docile as the other classes, yet they displayed great aptitude to fatten.

The dark-greys were frequently to be found intermixed with them in the same herd: they were so called from the broad white stripe which extended the whole length of the back, and also the parts usually now white on the different parts of the body being thickly interspersed with small red spots; their horns were rather shorter, and had a more upward tendency, they were also smaller in size and smoother in their hair than those of either of the other classes; better on the chine than the mottle-faces; they fed very even, and their flesh was of very excellent quality.

The mottle-faces were known as the "Tomkin's breed," although that eminent breeder always maintained that his bull Silver (41), which was a red-with-white-face, was the best stock-getter he ever had; and as he was bred by him in early life, he formed the foundation of his breeder's future eminence: thus it appears rather strange why he should have diverged from the red-with-white-face breed.

After the death of Mr. Tomkins, his herd was brought to the hammer in 1819, when twenty-eight breeding animals realized £4,255 11s., being an average of nearly £152 each; the price of each animal, and the names of the purchasers are given in the appendix to Vol. I.

Mr. Price, of Ryal, near Upton-on-Severn, was also a very celebrated breeder of this class, he having purchased his cows with Mr. Tomkins. I have before said challenges were frequent in those days between the admirers of rival breeds; and Mr. Price threw down the gauntlet in a challenge to show twenty of his breeding cows against the same number of the longhorn breed for a hundred guineas. This was in 1812, two years subsequent to the bull Crickneck's triumph; and Mr. Meek, of Lichfield, again came forward, accepted the challenge, and was again defeated. These struggles between longhorn and middlehorn tribes were, as you will observe, a few years subsequent to the death of the celebrated Bakewell, and at which period the longhorns maintained a high position. In the year 1839 we again find Mr. Price sent forth a challenge to show twenty of his cows and a bull, all of his own breeding, against the same number of any one person's breeding, and of any breed, open to all England for one month. This challenge remained unaccepted, though it gave rise to a rather angry correspondence with Mr. Bates, of Kirkleavington, in the columns of the *Mark Lane Express*. Mr. Price had several very excellent sales of stock, which although below the average realized at the dispersion of Mr. Tomkin's herd, yet, when the greatly-increased numbers are taken into consideration, they very well compare with it, and rank amongst the highest on record; the particulars of two of those sales are given in the appendix to Vol. I.

Amongst other eminent breeders of these classes were the Earl of Radnor, Lord Talbot, Sir F. Lawley, Sir F. Goodricke, the Rev. J. R. Smythies, Capt. Rayer, Capt. Peploe, Mr. Smith, Mr. Drake, and Mr. Jellicoe; at the present day I believe Capt. Peploe, of Garnston, Hereford, and Mr. Smith, of Shelsley Walsh, Worcester, are the principal and almost the only breeders of this once-fashionable blood. In the early days of the meetings of the Royal Agricultural Society of England, many prizes were won by animals from the herds of Mr. Price, the Rev. J. R. Smythies, Mr. Drake, and others; yet now we rarely see any of the class exhibited at these meetings. Mr. Smith did send this year some good-fleshed animals to Worcester, but was not successful. The last in the female classes which I remember to have seen a winner, was the heifer Superb,* exhibited by the Earl of Radnor at Salisbury, and then purchased for the Royal herd, when she was put to the red-with-white-faced bull Brecon (918), and brought the heavy-fleshed bull Maximus (1650). There is a beautifully-executed lithograph of him facing the title-page of Vol. IV.; he was a winner of the first prize in his class at the Warwick and Battersea meetings of the Royal Agricultural Society of England. You will see the marks upon his face show the transition

---

* Superb was a winner of the third prize in her class at Bingley Hall, and second at Islington this Christmas.

from the mottle-face to the red-with-white-face: they are larger than those of the mottle-faced, and fewer in number. At the Smithfield Club shows the steers and oxen were in by-gone days also very successful: at the same time I cannot help thinking that their success was subsequent to the year 1812, when the note was made respecting the red ring round the eye; perhaps that circumstance denoted the intermediate stage when the white face was being changed for the mottle. One of the most successful victories achieved by any breeder at the Smithfield show—and I think I may add of any breed—was the exhibition of three animals bred and fed by Mr. Hayton (afterwards Mr. Gwinnett), of Moreton, Hereford, with which he gained three first-prizes in the same year. The gold-medal steer of 1859, at Bingley Hall and Smithfield, bred and fed by Mr. Shirley, Bawcott Munslow, Salop, was the result of the use of a red-with-white-face bull upon a cow possessing a considerable portion of mottle-faced blood: he was 2 years 6 months 3 weeks 6 days old. When slaughtered his girth was 8 feet 7 inches, his live weight 15 cwt. 23 lbs.; dead weight 153 stones. The first seventeen months of his life he was treated in the usual way of rearing calves and steers, and therefore only confined to stall-feeding about thirteen months: it is worthy of remark that his dam, Silky, was also the dam of the gold-medal steer at Bingley Hall, Birmingham, in 1857. The weight and girth of this animal strikingly contrasts with that of Mr. Westcar's prize ox of 1799; but it must be borne in mind that the age of the one was six or seven years, whilst that of the other was a little over two and a half years.

The light-grey or white was held in great esteem with the red-with-white-face breeders, and were very frequently to be found amongst some of the best of their herds. They were a very noble class of animal; and, with the exception of the difference in colour, they closely assimilated with the red-with-white-face class. Many of Mr. Westcar's triumphs were won with light-grey or white oxen or steers. The following are some interesting particulars of one of them:—A 6 years old ox, bred by Mr. Tully, Huntington, fed by Mr. Westcar, and taken from a picture in the possession of the family of the late Mr. Tully: carcase 268 stone 6 lbs. (of course the London stone of eight pounds), hide 13 stone 7 lbs., head

8 stone, entrails 7 stone 5 lbs., pluck 6 stone 1½ lbs., fat 34 stone 3 lbs.*
Mr. Gibbs, in his history of the Club, gives the weight of Mr. Westcar's stall-fed ox in 1802: carcase 274 stone 6 lbs., and his grass-fed ox that year 225 stone 6 lbs. Some of the finest oxen I have ever seen were of this breed, and were well known as the Tully-greys. The late Mr. T. A. Knight, of Downton Castle, Ludlow, was for many years an admirer and very successful breeder of them; and through his liberality in giving the use of his bulls to his numerous tenantry, and in some cases even the bulls themselves, they became dispersed over a large district of country around his noble domain, and there they obtained the name of the "Knight-greys," and perhaps very properly so too, inasmuch as their general character differed in some essential points with the Tully-greys. They were smaller in size, more even and firmer in their flesh, and displayed an upward tendency of the horn, thus showing that an evident commingling of blood had taken place with the light and dark grey: this is also shown in the coloured engraving of Broxwood (485); and a reference to his pedigree will show, that although he was of a light-grey colour, yet he embraced some of the best red-with-white-face blood: his grt. grt, gr. sire, Conservative, was a mottle-face; he combined the whole of the four distinct classes; yet his dam being a white cow she gave him a position amongst the light greys.

Of the Tully-greys, in addition to Tully of Huntington, and Tully of Clyro, Jones of Breinton, and Yarworth of Brinsop, Hereford had some very superior animals of the blood. The Knight-grey breeders were Salway, Ricketts, Ashwood, Carter, Dawes, Longmore. It was at the Ashley Moor sale, when Mr. Salway's herd was dispersed, that the late Lord Berwick purchased the cows which laid the foundation of his Lordship's future eminence as a breeder.

Although the value of the light-greys as flesh-producing animals is admitted by all, yet they too have fallen (like the two preceding classes I have noticed) to the influence of the red-with-white-face bulls. Yet in

---

* I understand his sire was bred by Mr. Yeomans, and was red-with-white-face; but many of the calves by him were grey, which so annoyed Mr. Tully that he sent some to the butcher, and complained of the circumstance to the breeder, who fortunately induced him to rear those he had not sacrificed, and thus the origin of the Tully-greys.

their fall they have been productive of some of the choicest specimens of Herefords the world has ever seen. Amongst these I will mention Walford (871), winner of the first prize in his class at the Windsor meeting of the Royal Agricultural Society of England, and the gold medal at the Paris International show in 1855, as the best stock bull of any breed, besides many local prizes: also, Goldfinder (383) was not only a winner of many prizes, but proved one of the best stock-getters of his day: his sire was a grey. I could extend this list to great length; but I consider I have given you sufficient to prove the value of the light-grey class, and their intermingling of blood. The present proprietor of Downton Castle, Mr. A. R. Boughton Knight, has re-established a herd of them; he exhibited some choice specimens from it this year at the Worcester meeting of the Royal Agricultural Society of England, and won a second prize with a heifer-calf.

Having thus briefly noticed three out of the four classes into which Mr. Eyton saw it requisite to divide the Herefords, I now come to the class which appears on all sides to be the recognised race of cattle (now, as in the early days of the Smithfield Club) universally acknowledged as "The Herefords," and described in Eyton's Herd-Book as the "red-with-white-face." From the coloured engraving in Vol. I. you will see the characteristic marks of this class. They are:—The face, throat, chest, lower part of the body and legs, together with the crest or mane, and the tip of the tail, a beautifully clear white; a small red spot on the eye, and a round red spot on the throat, in the middle of the white, are distinctive marks which have many admirers. The horns are of a yellow or white waxy appearance, frequently darker at the ends; those of the bull should spring out straightly from a broad flat forehead, whilst those of the cows have a wave, and a slight upward tendency. The countenance is at once pleasant, cheerful, and open, presenting a placid appearance, denoting good temper, and that quietude of disposition which is so highly essential to the successful grazing of all ruminating animals; yet the eye is full and lively, the head small in comparison to the substance of the body. The muzzle white, and moderately fine, cheek thin. The chest deep and full. The bosom sufficiently prominent. The shoulder-bone thin, flat, and sloping towards the chine; well covered on the outside with mellow flesh; kernel full up from shoulder-point to throat; and so beautifully do the shoulder-blades bend into the body, that it is difficult to tell in a well-fed animal where they are set on. The chine and loin broad; hips long and moderately broad; legs straight and small. The rump forming a straight line with the back, and at a right angle with the thigh, which should be full of flesh down to the hocks, without exuberance, twist good, well filled up with flesh even with the thigh. The ribs should spring well and deep, level with shoulder-point; the flank full, and the whole carcase well and evenly covered with a rich mellow flesh, distinguishable by its yielding with a pleasing elasticity to the touch. The hide thick, yet mellow, and well-covered with soft glossy hair, having a tendency to curl. Such are the requisite characteristics of a first-class Hereford. The specimen selected for the embellishment of the first volume, to represent the red-with-white-face class, is Cotmore (376): he was the winner of the first prize in his class at the first meeting of the Royal Agricultural Society of England, held at Oxford: he was also a winner of many local prizes, and was perhaps one of the finest bulls ever seen; therefore very happily selected to embellish that volume: his colossal proportions were something very astounding, as may be inferred from the fact that the live-weight was 35 cwt. He was bred by Mr. Jeffries, the Grove, near Leominster. Sovereign (404) when at the age of fifteen years was his sire, but he was not of the same enormous size, although acknowledged to be one of the best stock-getters of his day. He was bred by Mr. Hewer, and very closely in-and-in bred, and used very extensively in different parts of the kingdom. I cannot close my remarks respecting this class without noticing another animal which, like his great grandsire Sovereign, was in-and-in bred, namely, Sir David (349), the former being by Favourite (442) from his own sister Countess, the latter being by Chance (348) from his own daughter. Yet the constitution and value of Sir David as a stock bull were something very remarkable: he was bred by Mr. Williams, Newton, near Brecon, calved in 1855, and after Mr. Williams' death purchased by Mr. Carpenter, Eardisland; he subsequently became the property of Mr. Price, Pembridge; and after winning the first prize in his class at the Newcastle and the Norwich meetings of the Royal Agricultural Society of England, besides several local prizes during the years 1847, 1848, and 1849, he was sold to Mr. Lumsden, Auchery House, Aberdeen, from whom he was purchased by Mr. Turner, Noke Court, and brought back not only to England, but to the very parish which he had for some three years left. There he became the sire of that celebrated stock-getter Sir Benjamin (1387); he was afterwards sold to Mr. Higgins, Woolaston Grange, Chepstow, and ended his career at Cronkhill, Salop, where he was sold by the late Lord Berwick to the butcher at the age of fifteen years. I have selected these as two very remarkable cases of in-and-in bred animals, each displaying great constitution, living to great ages, and as being the sires of many very first-class animals. But although this marked success attended their use, yet I consider too close an affinity of blood should as a rule be strictly guarded against, as its adoption much more frequently ends in disappointment, and consequently loss to those who try it. Amongst the most successful sales of this blood, may be noticed that of Mr. Williams, of Thingehill Court, Hereford, in the year 1814, when fifty-two breeding animals, including young calves, sold at an average of nearly £32. Mr. Hewer, Mr. Jeffries, Mr. Kedward, and Mr. Yeoman's sales were also of a high average.

Amongst the earliest breeders of the red-with-white-face class are the well-known names of Williams, Weyman, Hewer, Walker, Yeomans, Jeffries, Tully, Jones, Skyrme, Turner, Yarworth, Griffiths. Were I to attempt to enumerate the breeders of the present day, I should require more time than I am sure your patience would allot me; and were I to mention the names of some without giving that lengthened list, I should create a feeling the very opposite to my wishes. I will therefore refer you to the pages of the volumes I do myself the pleasure of presenting to you, to ascertain who they are, and where they are to be found. In those volumes I have given as careful information as I have been enabled to obtain, not only of the pedigrees of the animals, but also of the numerous national and local prizes won by them; and although I cannot say that information of that kind should be taken as an infallible guide in the selection of breeding animals, yet unquestionably the severe competition in the show-yard of the present day is such, that animals must possess considerable merit to be so placed; and therefore merit must of necessity be contained in the herds whence they emanate. At the same time I feel that the overfeeding requisite to obtain that object is an *evil*, and one which is not easily remedied, inasmuch as the breeder who has temerity enough to exhibit an animal in fair store condition is sure to stand but little chance of carrying away the prize for which it competes; and although not only the judges, the breeders, the public, and last, *not least*, that mighty engine for the dissemination of knowledge the press, condemn the practice, yet the fat animal catches the eye, and the others are designated an altogether middling lot. The result of this is, that their owners, believing in their superiority, resort to more forcing feeding, and upon subsequent competition they frequently become more successful: the judges place them first, the public admire them, and the press remark upon the wonderful improvement they have made since last they saw them; but unfortunately this is too often effected at a serious loss to the owners, from the destruction of the breeding properties of their animals with which they have ultimately become successful in obtaining the temporary advantage attached to a first-prize.

In the commencement of this paper I gave extracts showing the almost fabulous size attained by the Herefords exhibited in the early days of the Smithfield Club. In those days oxen were as highly esteemed in the county, the seat of their breed, for their working powers, as by the grazier for their meat-producing properties. Possessing as they did the weight of the shorthorn, with the activity of the Devon, they were very valuable for that purpose, being broken to the plough at two and under three years of age: they were kept by their breeders in the teams for two or perhaps three years; they were then sold to the grazier, and at about six or seven years old they found their way to Smithfield of the gigantic size I have mentioned. But the wants of the rapidly-increasing population of this nation have rendered a quick supply of meat requisite to meet the demand; and the great aptitude to fatten, and early maturity of the Herefords, admirably adapting them for that purpose, their working powers have been dispensed with, and they have gradually, and I may say almost imperceptibly, passed from their breeders year by year at an earlier age to the grazier; and thus we now rarely see them in the county exceeding three years old. Many of them have this year been sold from their pastures under that age, at prices varying from £25 to £30 each, without having tasted any kind of artificial food. Therefore, although we cannot now boast of our oxen of the present day, when they appear at the Smithfield Cattle Show, as standing 7 feet high, and girthing 12 feet 4 inches, yet we can and do pride ourselves upon their long cylindrical massive frames upon low legs, being evenly covered with heavy flesh of a girth surpassing that of most other breeds; and I consider the weight of Mr. Shirley's steer of 153 stones at 2 years 6 months old will very fairly compare with those oxen of Mr. Westcar's, which were 6 or 7 years old, and goes far to prove that although they do not now measure so much in height as they then did, yet that they would become of an equal or greater weight if kept to the same age—thus, no degeneration of their valuable properties are lost, but, on the contrary, a great improvement in quality and evenness of flesh has taken place. In proof of this opinion I could give many more examples, but I feel that in doing so I should only further exhaust your patience without any corresponding benefit.*

Of the milking properties of the Herefords, much of that depends upon the system adopted by the breeders in rearing the young, and can be greatly improved by proper treatment. I have before said the soil of Herefordshire is neither adapted for feeding or dairy, it being decidedly a breeding county, and little has been done by any to alter this state of things; therefore the progeny generally run with the dam. Sometimes a cow is taken to the pail, and the offspring, when either a steer or heifer calf, is put to assist another calf, and thus one cow rears two calves, whilst the other is taken to the pail. But in dairy counties, where the offspring are taken by hand, and the milking properties of the cow are well attended to, it has been done with the most satisfactory results, some of which are combined in the highly interesting information which I have obtained from several gentlemen of well known standing in the agricultural world, who have kindly favoured me with the result of their experience, not only in different parts of the United Kingdom, but also from Canada, Jamaica, America, and Australia.

---

* Since this paper was read, the gold medal has been won by a four-year-old Hereford ox at the Smithfield Cattle Show, bred by Mr. Meire, Cound Arbour, Salop, and fed by Mr. Heath, Ludham Hall, Norfolk. Girth 9 feet 4 inches. Length 5 feet 7 inches.

The Herefords having almost the exclusive possession of not only the county from whence they take their name, but also the counties of Monmouth, Brecon, Radnor, and Salop, I have not sought for information from those counties to lay before you, as I consider that fact is the best assurance of their value. With this feeling I extended my enquiries over a much larger area, and in districts where they have had to compete with other very valuable breeds, and I will now endeavour to lay before you the result of my enquiries in as concise a manner as possible.

The first I select is from this immediate neighbourhood, viz., Mr. Read, Elkstone. He writes me, that he finds the Herefords retain their general character and aptitude to fatten. That previous to his adopting cultivation by steam [some three years since] he kept several teams of working oxen; those of the Hereford breed he found best adapted to the purpose; having frequently tested them with the shorthorns, he unhesitatingly advances that they are more active, more hardy, more enduring, and better workers, and that such is the general experience of the district, that even where others are bred, the Herefords are purchased for that purpose. He finds that his herd generally endures the hardships of his cold wet district better than any other kind, and adds, "They have been used for dairy purposes for nearly half a century upon the farm," and that he believes they yield a larger return than could be obtained from any other breed upon a similar class of land. He rears his calves by hand after a few days old.

Mr. Bennett, North Cerney, in his letter, endorses Mr. Read's opinion, adding that during his thirty years' experience, he has tried some crossbred (Hereford and Shorthorns), but that he did not find them such valuable working oxen as the pure-bred Herefords; "they neither kept their flesh nor worked so well."

Mr. J. Barton, Manor House, Colne Fairford, writes to precisely the same effect, stating also, that he has been a breeder of Herefords ten years, that he has had a few good shorthorns, that he has bred them pure and crossed with the Herefords, but that his pure-bred Herefords are larger, better, and fresher than the others, although all live the same.

Mr. Lane, Compton Casey, Cheltenham, says, that he has been a breeder of Herefords sixteen years, that he finds they retain their general character, and his opinion is highly favourable to the working power of the Hereford. These are all opinions in your neighbourhood. I will now go to another county—Bedfordshire. Mr. Coleman, farm steward to His Grace the Duke of Bedford, writes me that in 1817 the herd at Woburn Park Farm consisted of 35 cows and heifers; some of them were purchased the previous year at Mr. Price's Ryal sale, and the herd was increased in 1819 by others from Mr. Tomkins's sale, Pyon; that several of the cows now in the herd are descended from those animals; that they are equal in size to any Herefords he has ever seen, but they lack a little in quality to some. He finds an introduction of fresh blood requisite. The only difficulty he experiences in the herd "is in keeping that hair which Hereford oxen are so famous for;" adding, " I attribute this to the dryness of the climate." He considers that no herd will fatten faster on grass than well-bred Herefords.

Major-General the Honourable A. N. Hood writes me, that the formation of the Royal herd was in 1855; that the animals then purchased became readily acclimatized; that they retain their general character, constitution, and aptitude to fatten; and adds, " I consider them quite equal to any other breed of cattle for early maturity." And Mr. Brebner, the farm manager, writes me to the same effect respecting them.

Mr. James Mappowder, Blandford, Dorset, says: "Our herd of Herefords have been established nearly 30 years, and so far from their being degenerated with us, they are much improved, and Hereford dairies are becoming very common in this county. In proof that they are good for milk with us, we let nearly 100 cows to dairy people, and if I buy one of any other breed to fill up the dairy, they always grumble, and would rather have one of our own bred heifers. Our system is—we let our cows at so much per year, finding them in land, and making the hay; the calves being reared by hand with skim milk and linseed until three months old, when we take to them, and allow a quarter's rent of the cow for the calf at that age; they are then turned into the pasture." He states, he has tried shorthorns and Devons with the Herefords; that he finds the latter fatten faster, and to be of a much hardier constitution than either, and therefore better adapted for grazing on thin cold soils; and adds, "The Herefords command the top price with the butcher."

Mr. Lobb, Lawhitton, Launceston, Cornwall, says: "The first fifteen years of my farming I kept North Devons—for the past twenty-seven years Herefords: I much prefer the latter, as they are much hardier. When breeding Devons I generally had two or three each year die with *scanter*,* a complaint to which I found the high-bred Devon subject; but since I have bred the white faces I have never lost one with that disease. I find the Herefords retain their general character as well as any other breed in this part of the kingdom, that they do not consume more food than the Devon, but feed faster, and attain a much greater weight at the same age. I attribute this to their more docile temper; my bulls are used to North Devon herds in my neighbourhood, and the cross-breds are much sought after by the graziers. The butchers like the cross bred or the pure bred Herefords equal to the Devons."

Mr. Olver, Penhallow, Grampound, Cornwall, in giving his twenty years' experience with them, says: "I find they retain their character as well in Cornwall as in their native county. I consider they are peculiarly adapted for this humid fickle climate, where Devons become small and delicate, and shorthorns grow bony and coarse. When grazed upon the granite moors I have invariably found them do better than the Devons when running together. I rear my calves on skim milk. I fatten my steers at from three to four years old, with roots and hay, or grass. When fat they weigh from 6 to 10 cwt., varying of course according to age (beef is sold by the cwt. in Cornwall). Hereford cows are generally said to be bad milkers; that is contrary to my experience, and I feel persuaded that when such is the case it does not arise from any constitutional defect, but rather from mismanagement in rearing, or a deficiency of the constituents essential to the production of milk in their food. My cow Patience, bred by Mr. J. Y. Cooke, Moreton House, Hereford, has this summer given 14 lbs. butter per week; and Blossom, bred by the late Mr. Long-

---

* A local term for a disease which appears to be prevalent in the West of England. The animals suffering are attacked with diarrhœa, and finally die in consumption.

more, Buckton, Salop, gave 22 quarts of milk, yielding 2½ lbs. butter per day."

Mr. Franks, Thong, Gravesend, writes: "I have been a feeder of Herefords without intermission 33 years, partly in Surrey, Herts, Cambridgeshire, and Kent. I have kept them on the same pastures with shorthorns, Devons, polled and horned Scots, the best of each kind I could procure, and the result has invariably been that the Herefords have got fit for the butcher first, and when in the stalls they will get fat on a less quantity of artificial food, and require less hay. With these results I prefer them, for either grass or stall feeding, to any other breed. I like to have them from three to four years old; the only drawback to them is, they get a little too fat with kind treatment, and the butchers, from March to August, do not like them so well as the other breeds; but from August to Christmas, if they are grass fed, those for the best trade in London like them better than they do any others. At that age I generally get them from 120 to 130 stone of 8 lbs., and with forcing to a much greater weight. I feed from forty to fifty annually." Leaving England I hear from Mr. Burman, agent to the Right Hon. Earl Lisburne, Crosswood Park, Cardiganshire: "I have had considerable experience with shorthorns, and am decidedly partial to them, but I must confess I do not think they would retain their character in a similar degree to the Herefords in this cold wet climate. I have frequently been surprised, and remarked to his Lordship upon their hardiness on the mountain farms of this county. Occasionally some of the tenants send a mongrel cow to his Lordship's bull, and if a bull calf it is sure to be kept as such. The offsprings of this class of animal I often meet upon the upland farms, and that even to the third and fourth generation, and I find them equally as hardy as the common country stock, and displaying their feeding qualities to a much greater extent. His Lordship's steers are put to fatten at $2\frac{3}{4}$ years old, and after about six or eight months' feeding, they average from ten to twelve scores per quarter; some kept on for Christmas fourteen scores and upwards:—they always sell at a better price per lb. than the native stock of the country. The breeding stock is much exposed to weather, yet it is always healthy and in good condition; therefore with all my prejudice for shorthorns, I should be afraid to treat them as the Herefords are here treated. Some people say the Herefords do not retain their character out of their own district; but, in my opinion, it is the treatment they get, and the neglect of obtaining fresh blood, and when that is the case what else can be expected?" Mr. Powell, Eglwysnunyd, Glamorganshire, writes me a similar satisfactory statement as the result of his 13 years' experience; and from Montgomeryshire, Mr. Williams, Farm Agent to John Naylor, Esq., Leighton Hall, says: "We have tried shorthorns, Scotch cross-bred and native cattle, and we find the Herefords superior to either, as they fatten in less time in the pastures and in the stalls. We usually feed our steers, and at about three years of age their average weight taken from the pastures is 10 to 12 scores per quarter. We have fed a few up to four years old, and find they then weigh from 16 to 20 scores per quarter. Our graziers purchase Herefords in preference to any other breeds for feeding purposes."

From Scotland Mr. Lumsden, Auchery House, Aberdeen, writes me: "I have been a breeder of Hereford cattle twenty-five years, and continue to do so, as I find they pay better for their keep than any other breed, or at any rate than those I have tried against them, viz., shorthorns and Aberdeens. The Herefords are hardy and well adapted for this northern climate, whilst the best shorthorns I could find were delicate, and frequently died. They can be brought to the highest condition with grass and turnips, whilst the others, to bring them to equal condition, must get oilcake or grain or both. I have tried different crosses, and have never been able to raise either the pure shorthorn, or the cross with the shorthorn and Aberdeen, above 216 st. of 8 lbs. at 4 years old, whilst I have raised the cross cow between the shorthorn and a Hereford bull to 238 stones."

From Ireland, Samuel Gilliland, Esq., Brookhall, Londonderry, writes me: "I consider the Herefords the best class of stock I can keep for the butcher; they are much hardier than the shorthorn, and more easily fattened than either the shorthorn, the Ayrshire, or the Irish. I have crossed the Ayrshire and the Irish with them to much advantage, particularly the Irish, which are greatly improved in their fattening properties by the cross. I have had several weigh over 20 cwt. gross under three years old; I have not used them much in the dairy, but whenever I have, the milk has been found to be of a very superior quality." Mr. Gilliland's farm agent, Mr. John Murrison, says: "After eight years' experience upon this estate I find the Herefords readily become acclimatized, and retain their general character; they fatten more readily than any other, and are always in a fit state for the butcher. Although all under my care are fed the same, making no difference as to breed, whether shorthorns, Devons, Ayrshire, or Irish, I always find the Herefords fatten more upon the same quantity and quality of food than the other breeds. I do not dairy except for private use, but I have found that eight quarts of the Hereford's milk are equal to 12 of the Ayrshire or Irish. I feed out the cows at ten or twelve years old, averaging 8 to 10 cwt.; steers from three to four years, averaging from 10 to 13 cwt. Our butchers prefer them by 10s. per cwt. to those of any other breed."

R. W. Reynall Esq., Killynan, Killucan, West-Meath, says: "The Herefords readily become acclimatized; in fact, they improve from the moment they arrive in Ireland, and retain their general character in every respect. I have fed them with other breeds, and find them easier to feed (particularly on grass), than any other kind. I generally work the steers and feed them at nine or ten years old. I have sold them at from £30 to £40 each to the butcher. Our butchers prefer Herefords to any other breed, and except the pure-breds, the half-breds, sell for more money to the graziers in our fairs. They have been bred in my family for fully a hundred years, and they are now fast increasing in Ireland: there are at least twenty Hereford bulls in this county at the present time."

I have now, I think, given you sufficient information respecting them from different parts of the United Kingdom, and will now lay before you the experience of several gentlemen from more distant lands.

F. W. Stone, Esq., Moreton Lodge, Guelph, Canada West, says: "I am an extensive breeder of Shorthorns, which breed I think very highly of; yet I trust I shall answer your enquiries without prejudice. From what I remembered of the Herefords in my youth, and seeing the poor animals exhibited here for two or three years as Herefords, I thought they bore a striking contrast to them, and upon my visiting the Royal Agricultural Society of England's show at Canterbury, I was so much pleased with those

I saw there, that I resolved to purchase some, and send out to let the people of Canada see what pure-bred Herefords were. I therefore commissioned my brother to purchase at the sales of Lord Bateman and Lord Berwick's herds. My herd now numbers twenty-three, are good specimens, and attract the attention of all who see them. They readily became acclimatized and retain their general character. I believe them preferable on the whole to other breeds as grazers. Those I have appear at all times fit for the butcher, and I should think they would be most profitable for the western prairies. I have not had any experience in stall feeding; but, during our long winters they seem to equal, if not to surpass, others in condition, and I think them as hardy as any breed, and very suitable animals for this climate, which is very changeable: sometimes in twenty-four hours it varies 30 to 40 degrees. Our cattle generally have to put up with it without any attention, excepting in very stormy weather in the autumn, when we put them up, and during the winter keep them in stables or yards. The Herefords stand these changes equal to any, and, I believe, will be of great service in crossing the stock here as they become known. They are not generally known here; but most people who have seen mine are very pleased with them, and I think they will be more appreciated."

J. Edwards, Esq., Knockalva, Ramble P.O., Jamaica, writes me: "The secretary of the Royal Agricultural Society of Jamaica, when about to visit England after many years' residence in this island, was instructed to make enquiries respecting the breed of cattle most adapted to the wants of this country, and he subsequently reported in favour of the Herefords. This was in 1844; and the following year two young bulls were sent here, and seven years afterwards two others, both of which died soon after landing; therefore there was no change of blood in this herd until 1858, when Sir Oliver (1732) and Malcolm (1646) were imported. But such was the change effected in our stock, that Mr. Trollope, in his 'Travels in Jamaica,' wrote: 'At Knockalva, I looked at Hereford cattle which I have rarely, if ever seen beaten at any agricultural show in England.' These were only half breds; still the cross is so direct, that the Hereford bull carries all before him, and many of our half-bred cattle you would scarcely suspect as being any other than pure breds. As regards their aptitude to fatten, there is no stock in this country to compare with them in that respect; and for early maturity they stand unrivalled; indeed, there are no breeds here that can compete with them. All our cattle are grass-fed and receive no artificial food. Our three-year-old steers weigh from 700 to 800 lbs. carcase weight. The temperature in the summer afternoon stands at 90 degrees in the shade, and in the winter (if such we can call our Christmas), I have never known it lower than 57 degrees. In Jamaica we require a breed of cattle possessing the following qualifications: good workers, hardy, and of great aptitude to fatten, and I fear no contradiction when I say, that no breed displays those qualifications in so eminent a degree as the Herefords."

From America, John Merryman, Esq., Hayfield Cockeysville, Maryland, writes: "I have seen the Herefords in the Northern parts of New York and Illinois, and as far south as Richmond; the climate at each point seemed to have no injurious effect upon them: I have also seen them three months after importation, and their health had been uninteruptedly good. I know of no instance when they have shown any ill effect from the change of climate: I feel convinced they retain their general aptitude to fatten in this country, and such is the published statement of Mr. John Johnson, of New York, who is one of the most experienced feeders in the United States. I have not had much experience in fattening steers, as previous to the war I kept my bull calves, and sold them at from six to twelve months old, at an average of 100 dollars each, therefore was not a feeder; but in May, 1862, I had on hand thirteen steers at from twelve to fourteen months old, when I sold them to Mr. Henry Carrell. He grazed them until the 1st December; they were then housed, and by the middle of April they had consumed ten bushels of mill feed, equal parts bran and ground corn, with a fair allowance of clover-hay, the object being merely to keep them during the winter in the condition they were when housed. Under the same roof he was feeding a lot of steers for market, their allowance was three times as much as the young Herefords. In April the butcher came to purchase the old steers; but insisted upon having five of the Herefords with them. Mr. Carrell purchased another lot of steers in April, and grazed with the remainder of the Herefords; in July the butcher again came, and he had to part with the Herefords, in order to make sale of the others. After keeping them only an average of fourteen months, their return was 270½ per cent. above cost price. Mr. Carrell has fed cross-breds and shorthorns, but never before fed Herefords; he is now willing to give one cent. per pound live weight more for Herefords, than any breed he has ever seen. The climate of Maryland varies as much as any part of the United States: the thermometer generally goes 10 degrees below zero, between the 15th December and 1st of February, but seldom lower; the heat sometimes reaches 90 degrees between the 15th of June and 1st of August, although I have known it as hot on the 2nd September. My herd consists of animals descended from the herds of the Rev. J. R. Smythies, Lord Berwick, and Mr. Longmore.

W. Dangar, Esq., Turunville, Hunter's River, Australia, writes me from Cornwall: Previous to my leaving the colony for England, I had not any pure-bred Herefords, my importations having been Shorthorns; but Mr. Hobler, who occupied an adjoining station to mine, had Herefords, and his bulls occasionally strayed amongst my cows; the result was, I had a good many white-faced cattle. Thus, it frequently happened that five or six of those cross-breds were amongst a draft lot of probably 150 to 200, and when the dealers were taken to them the white-faces were sure to be pointed out as the right sort, and the expressions made—'I wish all the lot were like those beasts! prime beasts! real plums!' I had opportunities of seeing them when slaughtered at Sydney, when the carcase butchers told me their quality was excellent, that they equalled the Shorthorns in their rough fat, and that they preferred the white-faced bullocks to those of any other breed. I also found they travelled the long distances from the station to Sydney—say 250 to 300 miles—better than the others; a less number became lame on the journey, which is, I consider, very greatly in their favour. From what I have seen in other herds, the Herefords readily become acclimatized, and fully retain their general character, displaying an equal or greater aptitude to fatten than those of the Shorthorns; and from these combined advantages, I lost no time, upon my arrival in England, in purchasing and exporting Hereford bulls. My nephew has now my station, and he has recently purchased a three-year-old Hereford bull, bred by Mr. Reynolds, of Tocal Patterson, Maitland—he was a winner of the first prize as a yearling, a two-year old, and a three-year-old successively at their agricultural show, and the judges pronounced him to be a better bull of the breed than any they had ever seen imported. The Herefords are now more numerous in the colony than the Devons, and nearly equal the Shorthorns in number."

Of their flesh, Mr. Rowlandson, in his prize report (before alluded to) says: "The flesh of the Hereford ox is superior to all other indigenous breeds, for that beautiful marbled appearance caused by the intermixture of fat and lean which is so much prized by the epicure." This fact goes far to account for the cause of the bad sale for Herefords in the London market during the summer months, as experienced by Mr. Franks—the meat being too rich, when well fattened, for the hot weather.

Of their aptitude to fatten in proportion to the food consumed there is an interesting experiment, which although made some thirty-five years ago, and given in "Youatt on Cattle," p. 34, may not be out of place to be briefly mentioned here, as it is strikingly confirmed, not only by the numerous practical opinions I have quoted to you, but also by a more recent experiment made by Mr. Moore at Coleshill, and published in the Journal of the Royal Agricultural Society, vol. xxxvii, p. 342, where, after giving the full results of his experiment, Mr. Moore says: "I think it proves this—that the Herefords take the lead in grazing." Youatt says: "Three Herefords and three Shorthorns were selected and put together in the straw yard on the 20th December, 1827, and were fed in the open yard at the rate of one bushel of turnips per day each, with straw only, until May 2nd, 1828, when their weights were taken, and they were sent to grass: the Herefords weighed 23 cwt. 2 qrs., the Shorthorns 27 cwt. On the 3rd November they were taken from grass and put into the stalls, when their weight was—Herefords 33 cwt., Shorthorns 38 cwt. 14 lbs. From that time until the 25th March the Herefords consumed 46,655 lbs. turnips, 5,065 lbs. hay; the Shorthorns 59,430 lbs. turnips, 6,779 lbs. hay. They then weighed—the Herefords 37 cwt. 14 lbs., the Shorthorns 43 cwt. 2 qrs., being an increase of weight from the first weighing of 2 cwt. 3 qrs. 14 lbs. in favour of the Shorthorns, but they had whilst in the stalls consumed 12,775 lbs. more turnips, and 1,714 lbs. more hay. On the 30th March they were all sold together at Smithfield, when the Shorthorns fetched £97, the Herefords £96, being an overplus of only £1 to pay for the enormous difference in the food consumed, and the greater price given on account of the heavier weight of the Shorthorns at the commencement of the experiment." Besides, there is one point unnoticed in Youatt, viz., the additional consumption of food from the 2nd of May to the 3rd of November, which it is only fair to presume was in the same proportion. If further proof were wanting that they are small consumers of food in proportion to the meat they make, I think the very form of the aged animals of the breed displays it; as, no matter to what age they are kept, they rarely exhibit any superabundance of offal.

I consider I have now shown sufficient in confirmation of the opinions I advanced at the commencement of this paper, viz.: that the Herefords, although an acknowledged aboriginal race of cattle indigenous to the soil of the county from whence they take their name, readily become acclimatized, and retain their general character, not only throughout the United Kingdom, but wherever they have been fairly tried in distant parts of the world; also, that they continue fully to retain their reputation, which has for ages past been accorded to them, for aptitude to fatten; that the quality of their meat is unsurpassed, if equalled; that it is duly appreciated wherever they have been tried; that, by proper management, their milking properties are good; that for early maturity and hardiness of constitution they are equal if not superior to any known breed; that they are a most valuable race of animals for their working powers when required; and that whenever they have been fairly tried, the quantity of meat they make, in proportion to the food consumed, is such that they can justly claim to rank amongst the most valuable class of animals known for the production of animal food, and therefore the most profitable breed of cattle for the grazier.

Printed by Rogerson and Tuxford, 246, Strand, London.

www.ingramcontent.com/pod-product-compliance
Lightning Source LLC
Chambersburg PA
CBHW062235220526
45471CB00009B/3492